The Acupuncture Cookbook

Jim Ventresca, Doctor of Oriental Medicine

AcuPractice Press
Copyright © 2012 by James Ventresca
Edited by Claudia Welch

ISBN: 978-1-105-41790-0

Preface

A cookbook is a good thing. It's a quick reference guide to help you create good results in the kitchen. This Acupuncture Cookbook can help you get good results in the clinic. "Cookbook Acupuncture" is a term I've heard since I began to study acupuncture in 1982. It is usually derided as an inferior approach to treatment. I disagree. In my experience, almost every practitioner I know, and teacher I've studied with, uses recipes, although they usually call them point prescriptions. Of course, most of them justify their recipes with theories, but the recipes persist, and they get handed down and around from one person to another. Here's how it goes:

"How do you treat stubborn low back pain?"

"Well, I always like to try such and such or if that doesn't work, I'll try this and that,"etc.

So why not just call it what it is, and get the information out to as many folks as possible?

Of course, differential diagnosis is important. Holistic treatment dictates that one address both the root causes and the branches of disorders. I assume that, as a healthcare provider, you already have at your disposal one or more means of treating the root causes of your patients' disorders. We could fill a small library of Traditional Chinese Medicine (TCM) with books on how to treat the roots. This book is designed to help you manage the branches.

Superior cookbooks go into great detail on all the theories and techniques associated with the recipes in the book. This isn't one of those types of cookbooks. It's more like a card file of recipes from your grandmother. Your grandmother would have assumed you already know the basics, yet she might still have written in a few a reminders such as, how to mix the pancake batter: "Mix only until the dry ingredients are moistened." Because I assume you have prior training, as well as access to acupuncture charts, books, apps, and notes to fill in any gaps, this Acupuncture Cookbook only touches on the theories and techniques, as gentler reminders.

I am a good cook. I've been cooking since I was a boy. Yet still, I find it helpful to keep a cookbook or two in my kitchen, and open them from time to time, for a quick reminder of how to make something I haven't cooked for a while, to learn how to make a dish I haven't tried before, or inspiration for something new. I hope you find this Acupuncture Cookbook serves a similar purpose in your clinic.

Acknowledgments

Since I began practicing and studying acupuncture and Oriental medicine many years ago, I've had many wonderful teachers and colleagues from many different acupuncture traditions. The techniques in this book are drawn from all of them. It would be difficult—if not impossible, to trace the origin of each of these techniques, but I am ever grateful for and to all the folks who helped me along the way. Most especially I would like to thank Dr. James Tin Yao So who inspired the first few generations of acupuncturists in the US, and started the New England School of Acupuncture; Don Halfkenny who taught me how to be practical, fair and reasonable in dispensing acupuncture healthcare; Kiko Matsumoto-Euler for teaching me how to reason within the boundaries of acupuncture; Jeffery Yuen for teaching me how to understand the underpinnings of this medicine; Skya and Anthony Abbate for giving me the opportunity to learn how to teach; my brother Dr. Chuck Ventresca for his example of complete integrity as a physician and his mastery of the science and art of healing; and Dr. Claudia Welch, a skilled healer, talented author, my editor, my inspiration, best friend, partner, and wife.

Introduction

State rules and regulations, or individual practitioners, may make clear distinctions between the terms, "acupuncture," "dry needling," "meridian therapy," or "trigger-point needling," but it's all the same to me, and probably to your patients too. Each of these methods involve placing needles into painful areas to relieve pain. They all require some knowledge and experience to get good results.

Whether you are a licensed acupuncturist, a doctor of chiropractic, medicine, naturopathy, osteopathy, physical therapy, or Oriental medicine, this book assumes that you are a licensed healthcare practitioner, with acupuncture or dry needling in your scope of practice, and that you have taken at least basic acupuncture training. In other words, I assume that you already know how to insert a needle, which points are contraindicated in what conditions, how to find most of the important acupuncture points, and that you are qualified to practice. Hopefully you also know how to arrive at a basic TCM diagnosis like: Liver Qi Stagnation, Qi and Blood Stagnation, or Spleen Qi Deficiency. That said, like all of us, you may still need an acupuncture chart to find the point you're looking for, and you may need a little reference material to hone your diagnosis on a particular patient. That's all part of the experience that comes with time. Every time you use a point or make a diagnosis, and see good results, you'll remember that point or diagnosis.

The issue of who gets to practice acupuncture and how much training is necessary can be a hotly debated one. Individual state laws allow various healthcare providers to practice acupuncture, or dry needling, or whatever you might call it, with certain minimum amounts of training. Because their laws allow them to employ these techniques, some health care providers naturally want to, so they look for training programs that meet their state's minimum requirements. I believe everyone who practices acupuncture should have the best training that can fit into the number of hours required by law. I put in a great deal of time debating acupuncture laws rules and regulations over the last 30 years, and I'm glad I had the opportunity to do so. However, now I view my role exclusively as an educator, and as such, I'll leave the politics to others and concentrate on providing the best training I can to all acupuncture providers.

The information in this book is based on 30 years of personal experience treating patients and teaching acupuncture. It's short, but direct and to the point. I really think it's my best stuff. I hope you find it helpful.

Edition Notes

This is the first edition of this book. In the future I hope to expand it to include more information on how point prescriptions are crafted and perhaps include different charts and other reference material. Sorry, there are no pictures. I haven't got the skill required to draw appropriate pictures of the meridians, acupuncture points, trigger points, etc., and I assume you have access to other books and charts for visual reference. If I happen to sell enough copies of this book to pay someone to do the drawings, or to buy the rights to some good ones, I'll put them in the next edition.

By the way, I'm already thinking about, and making notes for, the second edition. If you have any thoughts, comments, questions, or suggestions for helpful information to be included in the next edition, I'd love to hear from you. Send me an email: acupractice@gmail.com

This book is set up in a kind of an outline format for a couple of reasons. First, it's how I think. I've been teaching this material for many years, and this is how I put my thoughts down on (electronic) paper. Secondly, this format makes it easier to reference. Hopefully, once you've familiarized yourself with the book, you'll be able to find what you need quickly and easily.

One other thing: These are the meridian abbreviations I use in this book:

LU: Lung	LI: Large Intestine
SP: Spleen	ST: Stomach
HT: Heart	SI: Small Intestine
KD: Kidney	UB: Urinary Bladder
PC: Pericardium	SJ: San Jiao
	aka Triple Warmer/Energizer
LR: Liver	GB: Gall Bladder
DU: Du Mai	REN: Ren Mai
aka Governing Vessel (GV)	aka Conception Vessel (CV)

Reference Material

I realize that the clinical application of acupuncture is difficult to convey in a book. **AcuPractice™ Seminars** is the program I offer, for training healthcare professionals in acupuncture. Most of our students are chiropractors, but we get a number of medical doctors, osteopaths, naturopaths, physical therapists and nurses as well. If you need a training course, would like some additional training, or if you have a colleague who would like to study, please consider AcuPractice™ Seminars. It's a really good program, and you'll find yourself both competent and confident to practice acupuncture by the time you've finished the program. For dates, locations fees, hours, etc., please visit our website: **www.AcuPracticeSeminars.com.**

If you haven't read *The Web That Has No Weaver* by **Dr. Ted Kaptchuck,** I highly recommend you get a copy and read it. It will furnish you with most of the information you need to understand the basic theories of acupuncture and Oriental medicine. Even if you're not going to use all the information in the book, you should at least make sure you are familiar with it, if only to be able to speak knowledgeably on the subject.

You should have a good set of **Acupuncture Charts**. I recommend the kind that hang on your wall, so you have quick reference, and so your patients will become more interested in acupuncture. You can certainly use a set of charts in book form, or even something printed off the internet. However you should have some way to easily locate points when you want them.

In addition to Acupuncture Charts, a good set of **Trigger Point Charts** is really worth having. When you are treating pain, you will want to be able to quickly locate and treat the related trigger points. Of course you can simply palpate for the trigger points, but why reinvent the wheel every day? Most common trigger points have already been discovered. The charts are worth having. This is especially true for acupuncturists and others who haven't had direct training in trigger points. If a Chinese doctor from the 17th century had taken the time to catalog all the Ah Shi points, that were useful for treating pain, along with their referral pathways, those points would be taught in every acupuncture school in the world. Well, it so happened that they were catalogued in the mid to late 20th century, by Dr. Janet Travell, and we call them Trigger Points. I believe any acupuncturist would be well-served by spending a little time learning how to find and work with them.

In order to get a clear perspective on how Oriental medicine views and balances health through lifestyle and diet I highly recommend *Balance Your Hormones, Balance Your Life: Achieving Optimal Health and Wellness through Ayurveda, Chinese Medicine and Western Science* **by Dr. Claudia Welch.** This book is geared toward women's health, but is clinically applicable to all patients, men and women. And a worthwhile read for all healthcare practitioners.

AcuPractice™ Seminars offers a **National Board Review Class** each year, as a part of the full program we offer in various locations around the US. If you feel you could use a review of the basic information, it's not a bad way to refresh your understanding of basic theory, points, techniques, etc. Again, see our website or drop me an email for more info.

My favorite reference book for acupuncture points is *A Manual of Acupuncture* **by Peter Deadman & Mazin Al-Khafji with Kevin Baker.** It's a great textbook with all you could want to know about acupuncture points. **They also have a fantastic App** available for your pocket device.

The Basics

This section briefly introduces many clinically useful theories that form the basis of acupuncture and Oriental medicine. These theories can be successfully employed to enhance the practice of acupuncture with only a little effort spent in understanding them. These concepts are extremely important when treating internal disorders. They are less critical for neurological and musculoskeletal disorders, but can still be quite useful for crafting more effective acupuncture treatments for these maladies.

Yin and Yang

Yin and Yang are the most basic ideas underlying traditional Chinese thought and medicine. Volumes have been written on this subject, and it is a rich and interesting topic to discuss, with many implications that can be interpreted in all aspects of our existence. Yin and Yang encompass any opposite and complementary pairings that exist. Night and day, hot and cold, up and down, male and female, left and right, are but a few of the myriad of possible pairs we find in our universe. Here we will only concern ourselves with the aspects of Yin and Yang that directly apply to the practice of acupuncture.

There are three sets of opposites that describe Yin and Yang in most clinical applications. Respectively, they are **"Substance and Function," "Internal and External,"** and **"Cooling and Warming."**

Yin represents the **substance** and **substances** of the body, the **internal** areas of the body, and the energies that **cool** and **moisten** the body, and provide for **rest**.

Yang represents all **functional** aspects of the body, the body's **external** aspects, and the energies that **warm** and activate the body.

Meridians and Organs

The Meridians and Organs of acupuncture have some areas of overlap but, for our purposes, we will consider them separately. Organs and Meridians are classified as either Yin or Yang and they come in pairs.

When I discuss the Yin Meridians and Organs I will, for the most part, be addressing metabolic dysfunction, or internal disorders what I refer to as Yin disorders. Yang Meridians and Organs are generally used to address the Yang disorders, those that manifest more on the surface or exterior of the body, such as pain, tension, and tightness. There are a few exceptions to this rule and we will explore them as well.

This book will mainly addresses the Meridians that are concerned with musculoskeletal and neurological conditions. I briefly address the Organs under Internal Disorders.

The Substances

Qi (pronounced "chi") is energy. I could write an entire book about the details and intricacies of Qi, (or someone could) but not today. Suffice it to say that Qi is the energy that flows throughout an individual, and activates all life's processes. The main functions of Qi in an individual are the following:

      To Move, Transport and Activate: Any movement, be it muscular, circulatory, respiratory, cellular, peristalses, or otherwise depends on Qi. If the Qi becomes deficient or stagnant, movement is impaired and disharmonies result.

      To Warm and Protect: These functions mainly apply to the body's ability to feel warm and to its ability to fight off pathogenic influences like colds and flus.

      To Hold in Place: Organs, Fluids and Blood all need to be held in place. When they are not held in place edema, prolapsed organs, easy bruising, varicosities, hemorrhoids, bleeding and other disorders result.

Blood (Xue). Blood is the nutritive aspect of the body's energy (Qi). The main functions of Blood are to nourish, and moisten an individual, and to provide the ability to rest and recover from disease. When a patient is Blood deficient we may see signs of fatigue, dry skin & hair, dry eyes, and difficulty falling asleep. Blood also restores our strength after illness.

Jin-Ye is the term used to describe the fluids of the body. The Jin are the clear, watery, more Yang fluids such as tears and sweat. The Ye are the more Yin and thick fluids such as the internal moisture and synovial fluids.

Jing is the constitutional energy that we inherit from our parents. As we age we "use up" this energy. As the Jing is depleted over a normal lifetime the process of aging progresses. When the Jing is depleted too quickly, signs of early aging occur.

Shen is the Oriental medical term for the individual's consciousness, and as such, it dictates the patient's subjective experience of pain. I've found that in almost all cases of pain, calming the Shen is very useful. It will relax the patient and allow for the free flow of Qi and Xue in the body. If the Shen is not at peace, it can result in further imbalances that lead to increased stagnation. This increases the tension, which increases the stagnation, which increases the pain, which increases the loss of peace-of-mind, which increases the tension, snowballing into more pain, both physical and psychological. I often begin treatments by calming the Shen.

The Five Elements_____

Wood, Fire, Earth, Metal and Water - Like Yin and Yang, five element theory is another way of classifying phenomena. The Five Elements permeate traditional Chinese thought, and can be applied to almost everything. While this is an important facet of Chinese medicine, it's not absolutely necessary to employ it in order arrive at a satisfactory Traditional Chinese Medicine (TCM) diagnosis. With a few exceptions, I'll leave the Five Elements for the next edition.

Causes of Disease_____

External causes of disease are mainly colds and flus. These are classified as **Wind, Cold, Heat, Dryness, and Dampness.** Mostly they reflect the different ways upper respiratory infections (URIs) manifest. Wind is usually the earliest stage of a URI. Cold is recognized by the presence of chills and body ache. Heat is characterized by red face, and tongue, and sore throat. Dampness produces excess mucous, and dryness a dry cough. Of course these pathogens often combine in various ways.

We see these external pathogens in other areas as well. For instance, cold usually brings cramping pain and can manifest in other areas of the body. Cold can enter the lower part of the body (Lower Warmer) and cause cramping menstrual pain, or into the joints causing arthridities. [?] Heat can enter the skin causing burning sores like boils, and carbuncles. Wind has similar effects to wind in nature. Wind can cause tremors and sudden unpredicted movements, or even paralysis. Dampness too is similar to what we find in nature. Like a damp basement, it's hard to resolve. Pain and feelings of heaviness which linger and may be accompanied by swelling tend to be damp. Dampness can collect as excessive mucous seen in the Lungs, or stools. Overweight is also a sign of Dampness.

Internal causes of disease are the emotions. When any emotion is experienced inappropriately, it will effect the functioning of the Organs it is associated with. Whether one experiences or expresses an emotion too strongly, or insufficiently (repressing it) it will have consequences. Anger affects the Liver, Grief affects the Lungs, Joy affects the Heart, Worry and Over Thinking affect the Spleen, and Fear affects the Kidneys.

Miscellaneous causes of disease are the many and various influences we encounter in our lives. Getting hit by a bus, bitten by a snake, contacting environmental toxins, and eating inappropriately are all examples.

Points: The Raw Ingredients

Many of the most popular acupuncture points have multiple uses. Many of these points can be added to almost any treatment, for varied conditions. You will see these ubiquitous points show up in many different point prescriptions. Again, like in a cookbook, if you are in the baking section, you will see flour, sugar, salt, butter, etc. show up time and time again, but in different proportions and used in different ways to get various results. Similarly, you will see many of the same points again and again when treating similar problems like pain, or disorders of the head & neck, or digestive disorders.

Just as it is beyond the scope of any cookbook to cover all foods, it is beyond the scope of this book to cover all the points and their many uses. There are many good and exhaustive books on the acupuncture points, and it is not my aim to repeat that information here. To keep emphasis on the most clinically useful material, I focus on the pathways of the main acupuncture meridians, and describe the locations and use for the points I find most useful, and which I commonly use. I also discuss these points and point combinations in the treatment many disorders.

In general it is good to keep in mind that **Yin meridians** and their points are used mainly in relation to the Organs, and so **treat more internal disorders.** The **Yang meridians**, on the other hand, are used for more surface disorders, and will be employed principally in the **treatment of pain syndromes**, or other symptoms that appear along the path of the meridian itself.

I recognize that every practitioner of acupuncture has his or her own favorite points. Each individual knows that there are some points that he/she has found that work best. Of course my point choice is based on my experience. When I look at the categories of points, I find that the actions that are associated with them are more applicable to some points, within the category, than other points. So the following lists of points have been edited by my personal experience.

Let me just say a few words about "Points Below the Knees and Elbows." I like them. I find that the most powerful points on the body are found from the knees and the elbows down. I don't mean to indicate that the other body points are not powerful, they are, but I always seem to get better results with the arm and leg points. Now, please remember that I often use other points, like Front Mu and Back Shu points as well as Scalp and Ear points, but I generally use them secondarily to the Points Below the Knees and Elbows.

The lists of points below are not complete. This is not an oversight or a misprint. I have included only the points from each category that I have found to be particularly useful.

Classic Point Categories_____

This table contains most of the classic point categories, and is what one might use to study for a national exam in acupuncture. Each of these categories have specific uses.

YIN Meridian	Jing Well Wood	Ying Spring Fire	Shu Stream Earth	Jing River Metal	He Sea Water	Yuan Source	Xi Cleft	Luo	Back Shu	Front Mu
LU	11	10	9	8	5	9	6	7	UB 13	LU 1
SP	1	2	3	5	9	3	8	4	UB 20	LR 13
HT	9	8	7	4	3	7	6	5	UB 15	REN 14
KD	1	2	3	7	10	3	5	4	UB 23	GB 25
PC	9	8	7	5	3	7	4	6	UB 14	REN 17
LR	1	2	3	4	8	3	6	5	UB 18	LR 14

YANG Meridian	Jing Well Metal	Ying Spring Water	Shu Stream Wood	Jing River Fire	He Sea Earth	Yuan Source	Xi Cleft	Luo	Back Shu	Front Mu
LI	1	2	3	5	11	4	7	6	UB 25	ST 25
ST	45	44	43	41	36	42	34	40	UB 21	REN 12
SI	1	2	3	5	8	4	6	7	UB 27	REN 4
UB (BL)	67	66	65	60	40	64	63	58	UB 28	REN 3
SJ (TW)	1	2	3	6	10	4	7	5	UB 22	REN 5
GB	44	43	41	38	34	40	36	37	UB 19	GB 24

One might reasonably assume that all the points in each of the above listed categories are effective as categorized. However, in my experience, that's not always the case. The rest of this chapter contains commentaries on what I believe to be the more effective acupuncture points, drawn from the above categories together with other clinically important categories, including the Extraordinary Meridian Points, the Controlling Points and a few others.

Eight Extraordinary Meridians Master/Couple Points

These are listed first because I believe that these are the most powerful points on the human body. These points are the ones that affect some of the deepest and most primal energies of the body. If one looks at the actions and effects of these points and then couples them with the energetics of the Extraordinary Meridians they are truly "Extraordinary Points." I choose from them first in almost all of my treatments, whether I am treating pain or internal disorders. Also, I often simply choose from these "Extraordinary Points" for their actions, energetic and/or effects.

Master Point	Couple Point	Extraordinary Meridian
SI3	BL62	Du
Lu7	Ki6	Ren
GB41	TW5	Dai
Sp4	P6	Chong
BL62	SI3	Yang Chiao
Ki6	Lu7	Yin Chiao
TW5	GB41	Yang Wei
P6	Sp4	Yin Wei

Point	Action / Energetic / Effect
SI3	Any Back Pain, shoulder pain
Lu7	Any pain or discomfort in the Head and Neck. Use with LI 4 to bring down any excess from the Head and Neck.
GB41	Soothe the Liver and Gall Bladder and address any discomfort in the waist, hips, intestines, pelvis
Sp4	Any digestive disorder; any Lower abdominal discomfort
BL62	Any Back Pain
Ki6	To strengthen the Kidneys
SJ5	Calm the Spirit (Shen) Expel Pathogens
P6	Any discomfort in the abdomen; Calming the Spirit

Dr. Jim's Tai Ji Treatment_____

I'll admit that this isn't really a point category, but these four points are so powerful when used together, that I consider them a category in and of themselves. A Tai Ji (Supreme Ultimate) Treatment is one that accomplishes many objectives with many different patients, and is used often to balance a patient, before or in lieu of, focusing on individual complaints. This particular Tai Ji Treatment is my favorite method for Harmonizing the Liver and Heart, which can be a very powerful method for calming the Shen/Mind/Spirit and many physical energies in the body. It uses three Extraordinary Points and Liver 3. I use it regularly on patients to harmonize the Qi prior to beginning a more individualized treatment. I discovered this treatment quite by accident. I originally learned it (I thought) from my teacher, Kiko Matsumoto, and found myself using it more and more frequently on many different patients. The results were so profound that some 5 or 6 years later, while I was teaching at New England School of Acupuncture, I ran into Kiko and mentioned how useful this treatment was. I was surprised when Kiko informed me that I had gotten it wrong. I had "heard" LR 3, when she said "SP 4." After getting over my embarrassment, I realized how glad I was that I had heard wrong.

I know that over the years, a lot of my patients have benefited from my "mistake.".

Needle-on-the-LEFT	Needle-on-the-RIGHT
P6	LR3
GB41	SJ5

The Controlling Points

While not a classic category, these are points and combinations that are useful additions to any point prescription associated with the area in question. You may not find all of these listed in other books, but they all work well.

LI 4 & LR 3 For **Pain Anywhere** in the Body
LI 4 & LU 7 control the **Head, Neck, Face, & Mouth**
ST 36 controls the **Digestion & Qi**
P 6 controls the **Chest to Navel** Area
UB 40 controls the **Low Back**
SJ 5 controls the **Hand**
SJ 3 controls the **Ear**
LR 3 controls the **Liver**
GB 26 controls the **Hips & Lower Warmer**

Hua Tuo Jia Ji Points

(0.5 cun lateral to interspinous spaces) These points can influence any problem associated with the nerves exiting at the level of the spine where you find the point. Think about dermatomes as well as internal influences. They are also extremely effective at releasing the paraspinal muscles.

Yuan Source Points

The Source points of the Yin Meridians can always be added to a point prescription to affect the Organ being treated. I'm not sure the Yang Meridian Source points work all that well, but the Yin ones surely do. I believe that these points are probably the best points for affecting the basic Yin and Yang energies of the Organ, especially of the Yin Organs. When it comes to tonifying Qi, Xue, Yin or Yang I've found the Yuan Source Points to be more powerful than the Back Shu Points or Du Points, with the possible exceptions of UB 23 and Du4

Ht 7: A very good point for tonifying Heart Yin, however I often use PC 6 in it's place. I like Ht 7 for difficulty sleeping from Ht Yin Xu (Deficiency) and Heart Xue Xu. If the patient has dream disturbed sleep, I prefer PC 8.

Lu 9: tonifies the Qi of the Lungs. I often tell my students not to overlook the Lungs in cases of general Qi Xu. Remember the Lungs are the "Master of Qi." Without their proper functioning the Qi cannot be utilized.

Sp 3, Liv 3, and Kid 3: I use these points to tonify their respective Organs. In fact, I often use them together as an alternative to SP 6

Five Element Points_____
When considering points from the Five Element classification I find that the Fire points and the Water points are the only ones I regularly choose from, for their Element correlation. There are of course a number of effective ways to use the Five Element points to construct wonderful and effective treatments based on the Sheng (Creative) and Ko (Controlling) Cycles, but as I said earlier, that topic is beyond the scope of this edition. I use both the Fire and Water points to cool heat in their respective meridians or organs. When needling the Fire points I almost always obtain a strong stimulus to disperse heat.
Fire Points

LR 2: is quite effective for reducing heat in both the Liver and the Gall Bladder. I use LR 2 as one of the principal points when treating oral and/or genital herpes, as well as shingles (herpes zoster) or any other LR/GB meridian heat. Red burning eyes, and Headaches with associated heat, also respond well to treatment with LR 2. I locate this point just proximal to the margin of the web between the big toe and the second toe. Needle it at a 45° angle in the direction of LR 3.

     KD 2: is effective in treating burning urination and other manifestations heat in the lower warmer. It is also very effective in treating heat along the Kidney and Urinary Bladder meridians. Because of the close communication between the Kidneys and the Lungs in water metabolism, KD 2 can also be effective in treating hot skin conditions. I often combine it with LR 2.

     LU 10: This seems to me to be the quintessential Fire point. For all hot Lung and skin conditions, this is the point of choice. I often couple it with LU 11. I locate this point in an unconventional way, and find that it is a very effective location. This point is located in the center of the belly of the thenar eminence. The point is needled toward the metacarpal bone of the thumb. You will find that needling this point in this way will result in a strong stimulus, and good heat reduction.

Water Points

     LU 5: is good for treating heat in the lungs, especially when the heat is accompanied by cough.

     KD 10: I use this point for treating heat from KD Yin Xu.

Five Shu Points

I love the imagery created by the Five Shu Points:

> *The Qi lies deep but is assessable, like water in a well, at the Jing-Well.*
> *At the Ying-Spring the Qi bubbles to the surface.*
> *The Qi gathers and begins to flow at the Shu-Stream.*
> *Force gathers and the Qi flows with vitality through the Jing-River.*
> *The Qi is flowing and moving as well as integrating as it forms the He Sea.*

That said, I find that the Jing-Well points are the only ones I use for their Shu-point effects. But, I use them all the time.

When treating pain, I find that expressing a few drops of blood from the Jing-Well point of the effected meridian can be one of the most effective parts of the initial treatment.

Luo Points

The distribution of the effects of the Luo Points covers a multitude of areas of the body. I really only use P 6 and LR 5 for these purposes:

P 6 is good for chest pain and any type of restlessness.

LR 5 is a point I often use for pain and discomfort in the genitals, as well as dampness or heat in the lower warmer.

Meeting or Influential Points

The Qi Meets at CV 17: I use this point to move the Qi of the Chest and Upper Warmer. **Always needle this point from superior to inferior <u>at an oblique angle</u>.** **Strong stimulation of this point is forbidden.** I always needle it with a gentle stimulation, but I like to get the Qi to move slightly down toward the belly.

The Hollow Organs Meet at CV 12: This is a very good point to harmonize digestion and assimilation.

The Pulse Meets at Lu 9: Again, the Lungs are the Master of Qi. If the pulse is weak, consider the Lungs, and LU 9 is especially good for generating Qi, which in turn generates the pulse.

The Nerves and/or Tendons Meet at GB 34: It is a good point for treating tightness and tension in the muscles and tendons. It is especially useful for tension and tightness along the course of the Gall Bladder meridian and in the mid-to-upper back and neck. I locate and needle this point deeper, and in a slightly different direction, than most sources suggest. Begin at the junction of the heads of the tibia and the fibula. Palpate below the junction into the deep depression that is about 1.5 Cun distal to the junction. This is the insertion point. Using a needle that is long enough, insert at about a $30° - 45°$ angle so that the needle contacts the point, which lies under the junction of the heads of the femur and the tibia.

The Bones Meet at BL 11: Include it in treatments of broken bones, osteoporosis and osteoarthritis and other bone disorders.

The Back Shu (Associated) Points

While I don't use many of the Back Shu Points, those I use, I use regularly. They are sometimes the points I use when choosing points to treat the root of a dermatome. However, just as often as not, I will choose one of the Hua To Jia Ji Points rather than the Back Shu Points. The Hua To Jia Ji Points are found on line with the Back Shu points, but only 0.5cun from the inter-spinus space. Many practitioners use these in place of the Back Shu points because they can be needled perpendicular to the skin and much deeper than Back Shu Points, and they release the paraspinal muscles very well.

UB 11 the Back Shu Point of the Bones: As mentioned above, this point is very good for helping with bone-knitting after a break, or any other bone disorders.

UB 13 the Back Shu Point of the Lung: This point is very good for pain and congestion in the lungs. I have found cupping at UB 13 to be helpful in quelling asthma attacks and relieving shortness of breath.

UB 18 the Back Shu Point of the Liver: A good point for moving Liver Qi Stagnation. I mainly use it when there is tension, tightness and pain in the mid and upper back.

UB 23 the Back Shu Point of the Kidney: Very good point for low back pain due to Kidney Qi, Yin or Yang Xu (deficiency). It is especially effective when used with Du 4 and combined with Moxa.

UB 29 the Back Shu Point of the Sacrum: Another good point I commonly use for lower back pain.

The Front Mu (Alarm) Points_____

The Front Mu Points are points that are particularly well suited for treating the internal Organs. Like Back Shu Points, I don't use many of them, but the ones I use, I use often.

Ren 12 the Front Mu Point of the Stomach is very useful when treating Stomach disharmonies and pain.

Ren 17, the front Mu of the Pericardium, & Liver 14, the front Mu of the Liver, are both good points to treat when there is constriction, pain or discomfort in the chest or ribcage. I always combine them with Pericardium 6.

Stomach 25 the front Mu of the Large Intestine is useful when there is discomfort in the abdomen with constipation, diarrhea, or any gripping pain in the abdomen. Combine this point with Stomach 36, Pericardium 6, and Urinary Bladder 25.

Ren 3 the front Mu of the Bladder can be quite helpful in treating burning urination, frequent or difficult urination. Combine this point with Kidney 2 or 3, and Stomach 29 or 30.

Trigger Points aka Ah Shi Points_____

I believe that Ah Shi points (locally tender) and Trigger Points are the same phenomena.   Just like I'm using "acupuncture" to refer to all needling, I'll just use the term "Trigger Point" in this text to refer to all points found by palpation, including Ah Shi Points. As far back as the classics, these points were discussed and their use described. However, I will offer one piece of advice. Whenever possible, find a point that is a recognized acupuncture point or Trigger Point. This is not hard; careful palpation moving slowly out from the painful, along muscle and/or meridian pathways will usually reveal Trigger Points that have been previously recognized.   If you spend a bit of time pursuing these points you will be rewarded with better results then just needling the first tender point you find.

The Meridians and Their Major Points

Remember, in this edition, I'm only listing the most common points I use. There are a lot more good and useful points.

The first two meridians I will consider are, the Du Meridian, and the Ren Meridian. Technically they belong to the Eight Extraordinary Meridians, and are not main meridians. However, while they have special significance, they also function like main meridians.

You may notice that the Yang Meridians tend to treat pain along the pathway of the meridian, and Yin Meridians tend to treat more internal disorders. This is consistent with Chinese Medical Theories, and worth noting.

Du Meridian "Governing Vessel"

The pathway of the Du Meridian runs from the perineum, up through the middle of the spine, over the head, and ends at the upper lip. All points are on the posterior midline of the body. It is the most Yang meridian on the body. Since Yang Meridians are often used to treat disorders along the pathway of the meridian, the Du is important, for treating any and all disorders of the back and specifically the spine.

Point	Location	Main Uses
Du 1:	Midway between the tip of the coccyx and the anus.	Hemorrhoids; Rectal or Other Prolapse; Rectal Bleeding
Du 2:	On the hiatus of the sacrum.	Sacral and Coccyx Pain
Du 4:	With the patient in the prone position, it's in the deepest hollow in the low back. Below the spinous process of the 2nd lumbar vertebrae.	Controlling Point for Lower Back and KD; All Lower Back Problems, Especially Weakness; Kidney Deficiency
Du 14:	Below the spinous process of the 7th cervical vertebrae.	Controlling Point for Neck & Upper Back; Pain, Heat & Fevers in the Upper Body; Releases Exterior; Tonfies Wei Qi;
Du 20:	On the midline of the head, approximately on the midpoint of the line connecting the apexes of the two auricles	Any Excess Disorders of the Head; Any Prolapse In The Body; Connects with Brain; Clears the Mind; Headache
Du 25:	At the tip of the nose	Nasal/Sinus Congestion

Ren Meridian "Conception Vessel"

The pathway of the Ren Meridian runs from the perineum, up the midline of the front of the body, ending just under the lower lip. All its points are on the anterior midline of the body. It is also the most Yin meridian on the body. Since Yin Meridians are often used to treat internal disorders, the Ren is important, for treating many internal disorders and especially those associated with reproductive functions.

Point	Location	Main Uses
Ren 2	On the midpoint of the upper border of the symphisis pubis.	Controlling Point for All GYN and Urogenital Disorders
Ren 4	On the midline of the abdomen, 3 Cun below the umbilicus	Tonify the Kidneys; All GYN & Urinary Disorders; Tonifies Jing, Yang, Yin, & Qi
Ren 6	On the midline of the abdomen, 1.5 Cun below the umbilicus	Tonify the Spleen; Relieves Stagnation in Abdomen
Ren 8	In the center of the umbilicus	Strengthen Digestion; Diarrhea; Yang Collapse (No Needle - Moxa Only)
Ren 12	On the midline of the abdomen, 4 Cun above the umbilicus	Harmonize the Stomach; All Abdominal Problems
Ren 17	On the anterior midline, at level with the 4th intercostal space	Descend the Qi of the Chest
Ren 22	In the center of the suprasternal fossa	Throat and Swallowing Problems

Lung Meridian

The pathway of the Lung Meridian runs from the second intercostal space, 2/3 the distance from the middle of the sternum to the acromion process, down the anterior surface of the arm and ending at the proximal radial corner of the nail of the thumb. Since Yin Meridians are often used to treat internal disorders, the Lung Meridian is important, for treating many internal disorders associated with the Lungs. These include all respiratory disorders and symptoms associated with colds and flus.

Point	Location	Main Uses
LU 1	Lateral and superior to the sternum at the lateral side of the 1st intercostal space, 6 Cun lateral to the Ren Mai channel.	Mu / Alarm Point of The Lungs: All Lung Disorders
LU 5	On the cubital crease, on the radial side of tendon of m. biceps brachii, located with the elbow slightly flexed.	Strong, Productive Cough; Any Heat and/or Fullness in Lungs;
LU 7	Superior to the styloid process of the radius, 1.5 Cun above the transverse crease of the wrist	Master Point of the Ren Mai; Controlling Point For Neck & Throat; Circulates the Qi of The Lungs. Luo Point; Use with LI 4 for Headaches and OPIs
LU 9	At the radial end of the transverse crease of the wrist, in the depression on the lateral side of the radial artery.	Source Point; Tonifies the Lungs; All Deficient Lung Problems
LU 11	On the thumb, 0.1 Cun proximal to the radial corner of the nail.	Jing Well Point: Clears Heat From The Lungs (Sore Throat, Tonsillitis, Etc); Moxa Opposite Side For Nosebleeds

Large Intestine Meridian

The pathway of the Large Intestine Meridian runs from the proximal radial corner of the index finger along the arm on the lateral border of the radial bone to the top of the shoulder, crossing the SCM on the neck to the lower border of the ala nsai, on the opposite side. Since Yang Meridians are often used to treat disorders along the pathway of the meridian, the Large Intestine is important, for treating any and all disorders of arm, neck and head.

Point	Location	Main Uses
LI 1	On the radial side of the index finger, about 0.1 Cun proximal to the corner of the nail	Jing Well Point: acute pain on the meridian; Clears Heat from the Head and throat; (sore throat, tonsillitis, red burning eyes, etc)
LI 4	On the dorsum of the hand, between the 1st and 2nd metacarpal bones, approximately in the middle of the 2nd metacarpal bone on the radial side.	Controlling point for Head and Face: Main point for pain and other excesses in the head neck and arm. OPIs; **CONTRAINDICATED IN PREGNANCY**
LI 5	On the radial side of the wrist. When the thumb is tilted upward, it is in the depression between the tendons of muscle extensor pollicis longus and brevis.	Good local point for pain in the thumb and area

Point	Location	Main Uses
LI 11-12	Two points, best palpated for. When the elbow is flexed, these points are in and around the area between the depression at the lateral end of the transverse cubital crease and a spot superior to the lateral epicondyle of the humerus.	Fever; elbow pain
LI 14	Just superior to the lower end of the deltoid muscle.	Deltoid and shoulder pain
LI 15-	Anterior and inferior to the acromion, on the upper portion of the deltoid muscle.	Major points for shoulder joint pain
LI 16	When the arm is in full abduction, in the upper aspect of the shoulder, in the depression between the acromial extremity of the clavicle and scapular spine.	
LI 17-18	Two points best palpated for. On the lateral side of the neck, about level with the tip of the Adam's Apple, on the SCM.	Good local points for throat and neck problems
LI 20	In the nasolabial groove, at the level of the midpoint of the lateral border of the ala nasi.	any nose problem, including nasal or sinus congestion; often used with LI 4 and LU 7

Stomach Meridian

The pathway of the Stomach meridian runs from the middle of the lower border of the eye socket down to the outer corner of the mouth, then back to the mandible and up to the corner of the hairline. From there it proceeds down the front of the body along the mid-clavicular line over the ribcage, where it moves closer to the midline and down to the upper border of the pubic bone. From here it travels out to the leg and along a trajectory just lateral to the crest of the tibia, and on to the foot, ending at the proximal medial corner of the nail of second toe. Since Yang Meridians are often used to treat disorders along the pathway of the meridian, the Stomach meridian is important, for treating any and all disorders of and eye, head, teeth, and leg.

Stomach Meridian

Point	Location	Main Uses
ST 3	Directly below the center of the eye, at the level of the lower border of the ala nasi.	Sinus conditions; knee pain
ST 4	Lateral to the corner of the mouth, directly below ST 3.	Facial paralysis; trigeminal neuralgia; herpes; mouth ulcers; gum problems; tooth pain
ST 5	Anterior to the angle of the mandible, on the anterior border of the masseter muscle.	lower jaw toothache
ST 6	One finger-breadth anterior and superior to the lower angle of the mandible where the masseter attaches, at the prominence of the muscle when the teeth are clenched.	lower jaw toothache, Bruxism

Point	Location	Main Uses
ST 7	At the lower border of the zygomatic arch, in the depression anterior to the condyloid process of the mandible - located with mouth slightly slack.	TMJ; upper jaw toothache
Point	Location	Main Uses
ST 8	.5 Cun within the anterior hairline at the corner of the forehead, 4.5 Cun lateral to GV 24.	Frontal and band-like headaches
ST 25	2 Cun lateral to the center of the umbilicus.	any intestinal problems; front mu of LI;
ST 28	3 Cun below the umbilicus, 2 Cun lateral to CV 4	All GYN problems; damp heat in lower warmer; genital herpes; leucorrhoea.
ST 29	4 Cun below the umbilicus, 2 Cun lateral to CV 3	
ST 30	5 Cun below the umbilicus, 2 Cun lateral to CV 2.	Cold and/or blood stagnation in lower warmer;
ST 31	At the crossing point of the line drawn down from the ASIS and the line level with the lower border of the pubic symphisis, in the depression on the lateral side of sartorius, when the thigh is flexed.	Strengthens and moves the thigh and entire leg
ST 34	When the knee is flexed, point is 2 Cun above the laterosuperior border of the patella.	xi cleft; acute breast discomfort; knee pain/ weakness
ST 35	When the knee is flexed, the point is at the lower border of the patella, in the depression lateral to the patellar ligament.	knee joint problems
ST 36	3 Cun below ST 35, one finger breadth lateral to the anterior crest of the tibia	Controlling point for Digestion: All digestive problems; builds qi and blood; regulates digestion

Point	Location	Main Uses
ST 40	8 Cun superior to the external malleolus two fingers breadth lateral to the anterior crest of the tibia.	Helps resolve phlegm anywhere in the body
Point	Location	Main Uses
ST 41	On the dorsum of the foot, at the midpoint of the transverse crease of the ankle, in the depression between the tendons of muscle extensor digitorum longus and hallucis longus, approximately at the level of the tip of the external malleolus.	Pain/ weakness and stiffness in the foot and ankle.
ST 44	Proximal to the web margin between the 2nd and 3rd toes.	Clear heat from head, face, mouth and gums
ST 45	On the lateral side of the 2nd toe, .1 Cun proximal to the corner of the nail.	Jing Well Point: acute pain on the meridian; Clear heat from head, face, mouth and gums

Spleen Meridian

The pathway of the Spleen Meridian runs from the proximal medial corner of the nail of the great toe along the medial edge of the foot, up the leg following the posterior border of the tibia. At the waist it runs parallel to the midline, along the mid-clavicular line until it departs to end at the mid-axillary line midway, between the axilla and the free end of the 11th rib. Since Yin Meridians are often used to treat internal disorders, the Spleen Meridian is important for treating many internal disorders associated with the Spleen. These include all digestive disorders and symptoms associated with fatigue, and dampness.

Point	Location	Main Uses
SP 1	On the medial side of the great toe, 0.1 Cun proximal to the corner of the nail.	Jing Well Point: Moxa for bleeding from deficiencies
SP 3	Proximal and inferior to the head of the first metatarsal bone, at the junction of the red and white skin.	Source Point: Tonifies SP
SP 4	In the depression distal and inferior to the base of the first metatarsal bone, at the junction of the red and white skin.	Master point of the Chong Mai, All menstrual disorders; All gastric disorders; masses in the abdomen
SP 6	3 Cun above the tip of the medial malleolus, on the posterior border of the medial aspect of the tibia.	strengthens SP, KD, yin, qi and blood; All GYN; moves the lower abdomen; calms the mind; all reproductive issues (CONTRAINDICATED IF PATIENT IS PREGNANT)
SP 9	On the lower border of the condyle of the tibia, in the depression on the medial border of the tibia.	All damp disorders; A mild diuretic point
SP 21	On the mid-axillary line, midway between the axilla and the free end of the 11th rib.	whole body pain and/or weakness

Heart Meridian

The pathway of the Heart Meridian runs from the middle of the axilla, down the medial surface of the arm and ending at the proximal radial corner of the nail of the pinky finger. Since Yin Meridians are often used to treat internal disorders, the Heart Meridian is important for treating disorders associated with the Heart. These include all mental / emotional disorders and symptoms associated with sleep.

Point	Location	Main Uses
HT 1	When the arm is abducted, the point is in the center of the axilla, on the medial side of the axillary artery.	Seldom used. Deep needling can access m. subscapularis
HT 7	At the ulnar end of the transverse crease of the wrist, in the depression on the radial side of the tendon of muscle flexor carpi ulnaris.	Source Point: Most commonly used HT point all heart disorders. Nourishes the HT, Calms the Shen; relieves insomnia
HT 8	When the palm faces upward, the point is between the 4th and 5th metacarpal bones. When a fist is made, the point is where the tip of the little finger rests.	tachycardia; heat in the Heart; dream disturbed sleep, night terrors
HT 9	On the radial side of the little finger, .1 Cun proximal to the corner of the nail.	Jing Well Point: extreme heat in the Heart; loss of consciousness; tachycardia

Small Intestine Meridian

The pathway of the Small Intestine Meridian runs from the proximal ulnar corner of the pinky finger, along the arm on the lateral border of the ulnar bone, to the back of the shoulder, through the scapula, up to the back of the neck past the ear, and ending lateral to the eye. Since Yang Meridians are often used to treat disorders along the pathway of the meridian, the Small Intestine is important for treating any and all disorders along its pathway, especially around the scapula and neck.

Point	Location	Main Uses
SI 1	On the ulnar side of the little finger, about .1 Cun proximal to the corner of the nail.	Jing Well Point: acute pain on the meridian; mastitis; insufficient lactation
SI 3	When a loose fist is made, the point is on the ulnar side of the hand, proximal to the 5th MP joint, at the end of the transverse crease at the junction of the red and white and skin.	Controlling point for Spine: Master Point of the DU Mai: stiff neck; all back pain; stroke, MS, Turette's and other wind disorders
SI 6	When the palm faces the chest, the point is in the bony cleft on the radial side of the styloid process of the ulna.	xi cleft; pain in the meridian; stiff neck; Shoulder pain
SI 8	When the elbow is flexed, the point is located in the depression between the olecranon of the ulna and the medial epicondyle of the humerus.	local point for elbow pain/ stiffness
SI 9	Posterior and inferior to the shoulder joint. 1 Cun above the posterior end of the axillary fold.	local points for shoulder pain and Range of Motion
SI 10	Directly above SI 9, in the depression inferior to the scapular spine.	

Point	Location	Main Uses
SI 11	In the infrascapular fossa, at the junction of the upper and middle third of the distance between the lower border of the scapular spine and the inferior angle of the scapula.	Most important local point for the upper back, and scapular area
SI 17	Posterior to the angle of the mandible, in the depression on the anterior border of the SCM.	local point for jaw, neck, and ear
SI 19	Anterior to the tragus and posterior to the condyloid process of the mandible, in the depression formed when the mouth is open.	benefits the hearing; jaw problems

Urinary Bladder Meridian

The pathway of the Urinary Bladder Meridian runs from the medial inner canthus up over the head just lateral to the midline, down the side of the neck where it separates into two pathways, both running parallel to the midline of the back. One pathway is located at the distance of the medial border of the scapula, and the other is halfway between the medial border of the scapula and the midline. At the sacrum the meridian moves out to the buttocks and down to the center of popliteal crease, and throughout the back of the calf, to the ankle where it runs just under the lateral maleoleous ending at the lateral proximal corner of the little toe. Since Yang Meridians are often used to treat disorders along the pathway of the meridian, the Urinary Bladder is important, for treating any and all disorders of back, and legs. It should also be noted that this meridian contains the "Back Shu Points." These are specific points for each of the Organs.

Point	Location	Main Uses
UB 1	0.1 Cun superior and slightly medial to the inner canthus	Main Point for All eye problems
UB 2	on the medial end of the eyebrow, or on the supraorbital notch	Alternate point for all eye problems
UB 10	1.3 Cun lateral to GV 15, in the depression on the lateral aspect of the trapezius muscle	All neck pain, weakness and stiffness
UB 11	1.5 Cun lateral to GV 13, at the level of the lower border of the spinous process of T1	Back Shu Point of the Bones: All bone problems; arthritis

UB 12	1.5 Cun lateral to the GV meridian, at the level of the lower border of the spinous process of T2	OPIs; headache; cough; stimulates wei qi
Point	Location	Main Uses
UB 13	1.5 Cun lateral to GV 12, at the level of the lower border of the spinous process of T3	Back Shu Point of the LU: all lung problems; builds wei qi; chronic or acute
UB 14	1.5 Cun lateral to the GV meridian, at level with the lower border of the spinous process of T4	Local Point for Upper Back Pain
UB 15	1.5 Cun lateral to the GV 11, at the level of the lower border of the spinous process of T5	Back Shu Point of the HT: all heart problems
UB 16	1.5 Cun lateral to GV 10, at the level of the lower border of the spinous process of T6	Local Point for Upper Back Pain
UB 17	1.5 Cun lateral to GV 9, at the level of the lower border of the spinous process of T7	Back Shu Point of the Blood: All blood problems; tonifies blood; skin problems from heat in blood; Back Shu Point of the Diaphragm: Hiccough; Hiatial Hernia
UB 18	1.5 Cun lateral to GV 8, at the level of the lower border of the spinous process of T9	Back Shu Point of the LR: benefits all aspects of the liver; smooths, harmonizes, and tonifies liver
UB 19	1.5 Cun lateral to GV 7, at the level of the lower border of the spinous process of T10	Back Shu Point of the GB: all GB problems

Point	Location	Main Uses
UB 20	1.5 Cun lateral to GV 6, at the level of the lower border of the spinous process of T11	Back Shu Point of the SP: all SP problems; benefits all aspects of spleen
Point	Location	Main Uses
UB 21	1.5 Cun lateral to the GV meridian , at the level of the lower border of the spinous process of T12	Back Shu Point of the ST: benefits all aspects of the ST
UB 22	1.5 Cun lateral to the GV 5, at the level of the lower border of the spinous process of L1.	Back Shu Point of the SJ: regulates and transforms fluids;
UB 23	1.5 Cun lateral to GV 4, at the level of the lower border of the spinous process of the L2.	Back Shu Point of the KD: all KD pathologies; regulates all aspects of the KD
UB 24	1.5 Cun lateral to the GV meridian, at the level of the the lower border of the spinous process of L3.	Local Point for Low Back Pain
UB 25	1.5 Cun lateral to GV 3, at the level of the lower border of the spinous process of L4.	Back Shu Point of the LI: low back pain; colon problems; constipation, diarrhea, etc.
UB 26	1.5 Cun lateral to the GV meridian, at the level of the lower border of the spinous process of L5.	Local Point for Low Back Pain
UB 27	1.5 Cun lateral to the GV meridian, at the level of the lower border of the 1st posterior sacral foramen.	Back Shu Point of the SI: used with CV 3 for damp heat in urine; sacroiliac joint problems

Point	Location	Main Uses
UB 28	1.5 Cun lateral to the GV meridian, at the level of the 2nd posterior sacral foramen.	Back Shu Point of the UB: all UB Problems
UB 29	1.5 Cun lateral to the GV meridian, at the level of the 3rd posterior sacral foramen.	Local Point for Low Back Pain
UB 30	1.5 Cun lateral to the GV meridian, at the level of the 4th posterior sacral foramen.	Local Point for Low Back Pain
UB 31	In the 1st posterior sacral foramen.	These are the 8 Liao Points: Used for all genito-urinary, GYN, low back & leg problems
UB 32	In the 2nd posterior sacral foramen.	
UB 33	In the 3rd posterior sacral foramen.	
UB 34	In the 4th posterior sacral foramen	
UB 35	On either side of the tip of the coccyx, .5 Cun lateral to the GV meridian.	Coccygeal pain
UB 40 (used to be #54)	Midpoint of the transverse crease of the popliteal fossa, between the tendons of muscle biceps femoris and muscle semitendinosis.	Controlling point for the low back: All Low back pain; clears heat

Point	Location	Main Uses
UB 57	Directly below the belly of muscle gastrocnemius, on a line joining BL 40 and tendo-calcaneus, about 8 Cun below BL 40.	Upper back pain/tension; used with UB 67 for spasms of the calf;
UB 60	In the depression between the external malleolus and tendo calcaneus.	occipital headache; neck pain; distal point for sciatica; for labor pain; "aspirin point"
Point	Location	Main Uses
UB 62	In the depression directly below the external malleolus.	Master Point of the Yang Qiao Mai: Lateral musculoskeletal problems
UB 67	On the lateral side of the small toe, .1 Cun proximal to the corner of the nail	Jing Well Point: acute pain on the meridian; turning a breech fetus; used with UB 57 for spasms of the calf; back pain; tight hamstrings

Kidney Meridian

The pathway of the Kidney Meridian runs from the center of the ball of the foot, along the medial edge of the foot, under the medial maleolus, up the medial aspect of the leg posterior to the tibia. At the waist it runs parallel--and just lateral to--the midline, ending at the top of the sternum. The Kidney Meridian is important for treating many internal disorders. These include all growth development and aging, reproductive and urinary disorders, and symptoms associated with fatigue and general weakness.

Point	Location	Main Uses
KD 1	On the sole, in the depression when the foot is in plantar flexion, approximately at the junction of the anterior third and posterior 2/3.	emergency point; loss of consciousness; brings heat down fro the upper body
KD 2	Anterior and inferior to the medial malleolus, in the depression of the lower border of the tuberosity of the navicular bone.	Heat/Fire in the throat and lower warmer; UTIs
KD 3	In the depression between the medial malleolus and tendo calcaneus, at the level of the tip of the medial malleolus.	Source Point; All KD disorders; weakness in the knees; low back; lowered libido, E.D.
KD 6	In the depression of the lower border of the medial malleolus, or 1 Cun below the medial malleolus.	master point of yin qiao; Tonifies KD yin
KD 7	2 Cun directly above KD 3, on the anterior border of tendo calcaneus.	Tonifies KD Yin and Vaporizes fluids
KD 11-21	0.5 Cun lateral to Ren and level with the Ren points. KD 11 lateral to Ren 2 on the superior border of the symphysis pubis. KD 21 under the sternum	reinforce the Ren points
KD 16	.5 Cun lateral to the umbilicus, level with CV 8.	Strengthens the KD
KD 22- 27	These points run from the bottom of the sternum to the depression on the lower border of the clavicle, and are all 2 Cun lateral to the Ren meridian.	These points soothe the Spirit, and should be palpapated for tenderness

Pericardium Meridian

The pathway of the Pericardium Meridian runs from the chest, down the medial surface of the arm, and ends at the tip of the middle finger. Since Yin Meridians are often used to treat internal disorders, and the Pericardium is closely related to the Heart Meridian, it is most commonly used to treat disorders associated with the Heart. These include all mental / emotional disorders and symptoms associated with sleep, as well as organic heart disorders.

Point	Location	Main Uses
PC 4	5 Cun above the transverse crease of the wrist, on the line connecting PC 3 and PC 7, between the tendons of palmaris longus and flexor carpi radialis	xi cleft point; chest pain; pain in the meridian
PC 5	3 Cun above the transverse crease of the wrist between the tendons of palmaris longus and flexor carpi radialis	calms the spirit; regulates Heart rhythm
PC 6	2 Cun above the transverse crease of the wrist between the tendons of palmaris longus and flexor carpi radialis	Master Point of the Yin Wei Mai: Controlling point for the Chest and Abdomen; calms the spirit
PC 7	In the middle of the transverse crease of the wrist, between the tendons of muscle palmaris longus and flexor carpi radialis.	Source Point: calms the spirit; regulates Heart rhythm
PC 8	On the transverse crease of the palm, between the 2nd and 3rd metacarpal bones. When the fist is clenched, the point is just below the tip of the middle finger.	tachycardia; heat in the Heart; dream disturbed sleep
PC 9	In the center of the tip of the middle finger.	Jing Well Point: extreme heat in the Heart; loss of consciousness; tachycardia; night terrors

San Jiao Meridian

However you choose to translate it, the pathway of the San Jiao is similar to that of the Small Intestine, but more radial. It runs from the proximal ulnar corner of the ring finger along the arm on the lateral border of the ulnar bone to the back of the shoulder, along the top posterior aspect of the trapezius, up to the back of the neck, around the ear, and ends just anterior to the tragus of the ear. Since Yang Meridians are often used to treat disorders along the pathway of the meridian, the San Jiao is important for treating any and all disorders along its pathway, especially around the shoulder, and ear.

Point	Location	Main Uses
SJ 1	On the lateral side of the ring finger, about 0.1 Cun proximal to the corner of the nail.	Jing Well Point; acute pain on the meridian
SJ 3	On the dorsum of the hand between the 4th and 5th metacarpal bones, in the depression proximal to the metacarpophalangeal joint.	Controlling Point for the Ear
SJ 4	On the transverse crease of the dorsum of the wrist, in the depression lateral to the tendon of muscle extensor digitorum communis.	Good local point for the wrist
SJ 5	2 Cun above TW 4, between the radius and the ulna.	Master Point of the Yang Wei Mai: local point

SJ 10	When the elbow is flexed, the point is in the depression about 1 Cun superior to the olecranon.	Good Local Point for the elbow
Point	Location	Main Uses
SJ 14 (posteri or to LI 15)	Posterior and inferior to the acromion, in the depression about 1 Cun posterior to LI 15 when the arm is abducted.	Shoulder joint pain and ROM
SJ 15	About 1 Cun posterior to GB 21. Midway between GB 21 and SI 13, on the superior angle of the scapula.	Tightness, tension and pain in the neck and shoulders
SJ 17	Posterior and superior to the angle of the mandible. Posterior to the lobule of the ear, in the depression between the mandible and the mastoid process.	All Ear disorders
SJ 21	In the depression anterior to the supratragic notch and slightly superior to the condyloid process of the mandible. The point is located with the mouth slack.	All Ear and Jaw disorders
SJ 23	At the lateral end of the eyebrow	All disorders involving the side of the head/face; lateral headaches, eye pain, ear pain

Gall Bladder Meridian

The pathway of the Gall Bladder Meridian runs from the Lateral outer canthus, back and forth across the sides of the head, down the side of the neck, where it follows the top of the trapezius, down to the side of the ribcage, throughout the flanks, to the hip, and down along the most lateral aspect of the leg, to the ankle, where it runs under the lateral maleoleous and ends at the lateral proximal corner of the fourth toe. Since Yang Meridians are often used to treat disorders along the pathway of the meridian, the Gall Bladder is important for treating any and all disorders of sides of the head, trunk, hip, and legs.

Point	Location	Main Uses
GB 1	0.5 Cun lateral to the outer canthus, in the depression on the lateral side of the orbit.	Secondary point for eye problems
GB 2	Anterior to the intertragic notch, at the posterior border of the condyloid process of the mandible. The point is located with the mouth open.	Good local point for ear and jaw
GB 8	Superior to the apex of the auricle, 1.5 Cun within the hairline.	Lateral Headaches; post stroke speech disorders; enters the brain
GB 14	On the forehead, 1 Cun directly above the midpoint of the eyebrow.	All eye problems; frontal and temporal HA
GB 20	In the depression between the upper portion of the SCM and the trapezius, just below the occiput.	All Wind: internal LV wind and external OPI wind; opens the orifices of the head: All disorders effecting the eyes, ears, and nose; All Headaches, especially occipital

Point	Location	Main Uses
GB 21	Midway between GV 14 and the acromion, at the highest point of the shoulder.	Primary point for neck and shoulder tension, pain and tightness. (CONTRAINDICATED IF PATIENT IS PREGNANT, OR HAS A HEART CONDITION)
GB 25	On the lateral side of the abdomen, on the lower border of the free end of the 12th rib.	Front mu of KD; pain in lumbar region
GB 26	Directly below the free end of the 11th rib, where the LV 13 is located, at the level of the umbilicus.	Main Point on the Dai Mai; All GYN Disorders
GB 28	Anterior and inferior to the ASIS, 0.5 Cun anterior and inferior to GB 27.	Secondary Point on the Dai Mai; All GYN Disorders
GB 29	IN the depression of the midpoint between the ASIS and the great trochanter. When locating this point, put patient in lateral recumbent position with thigh and knee both flexed to about 90^0.	Main Points for Sciatica, hip; lumbar to thigh, leg pain and paralysis
GB 30	At the junction of the lateral 1/3 and medial 2/3 of the distance between the greater trochanter and the hiatus of the sacrum. When locating this point, put patient in lateral recumbent position with thigh and knee both flexed to about 90^0.	

Point	Location	Main Uses
GB 31	On the midline of the lateral aspect of the thigh, 7 Cun above the transverse political crease. When the patient is standing erect with hands at sides, the point is where the tip of the middle finger touches.	Main Points for Sciatica, lumbar to thigh, leg pain and paralysis
GB 34	In the depression anterior and in inferior to the head of the fibula.	relaxes the tendons; good point for knee, sciatica and leg pain; pain anywhere in the body; shoulder pain
GB 40	Anterior and inferior to the lateral malleolus, in the depression on the lateral side of the tendon of m. extensor digitorum longus.	Ankle Pain
GB 41	In the depression distal to the junction of the 4th and 5th metatarsal bones, on the lateral side of the tendon of m. extensor digiti minimi of the foot.	Master Point of the Dai Mai: All GYN and LV disorders
GB 44	On the lateral side of the 4th toe, about .1 Cun proximal to the corner of the nail.	Jing Well Point; acute pain on the meridian

Liver Meridian

The pathway of the Liver Meridian runs from the proximal lateral corner of the nail of the great toe, along the medial edge of the foot, up the leg, posterior to the border of the tibia, between the Spleen and Kidney meridians. At the waist it runs parallel to the midline until it departs to the free end of the 11th rib, ending at the sixth intercostal space on the mid-clavicular line. Since Yin Meridians are often used to treat internal disorders, the Liver Meridian is important for treating many internal disorders associated with the Liver. These include many emotional disorders and symptoms associated with stagnation and wind.

Point	Location	Main Uses
LR 1	On the lateral side of the great toe, 0.1 Cun proximal to the corner of the nail.	Jing Well Point: restores consciousness; Inguinal pain; groin pulls; moves Qi in the lower warmer; genitourinary issues; stops bleeding
LR 2	On the dorsum of the foot, between the 1st and 2nd toes, proximal to the margin of the web.	LV fire and heat in the head; Heat in the Lower Warmer; Burning urination; red burning eyes; LV yang rising HA; Herpes I & II; Shingles; LV wind
LR 3	On the dorsum of the foot, in the depression distal to the junction of the 1st and 2nd metatarsal bones.	Source Point: Primary point to benefit all aspects of LV. Move LR QI, Quell LR Wind, etc.
LR 13	On the lateral side of the abdomen, below the free end of the 11th floating rib.	Front MU Point of the SP: Liver invading Spleen
LR 14	Directly below the nipple, in the 6th intercostal space.	Pain and tightness/fullness in the chest, ribs and breast

Extraordinary Points

Point	Helpful Translations	Location	Main Uses
Bi Tong	Nose Opening	At the highest point of the nasolabial groove.	Stuffy-Runny Nose
Jian Nie Ling	Shoulder Out Front	Midway between the end of the anterior axillary fold and LI 15	
Tai Yang	Great Yang	Draw a line following path of lateral end of eyebrow down to intersection of line from outer canthus of eye, at intersection, in depression.	Eye Pain / Discomfort Headaches
Xi Yan	Calf's Nose	A pair of points in the two depressions, medial and lateral to the patellar ligament, locating the point with the knee flexed. Lateral Xiyan overlaps with S 35.	Knee Pain
Yao Tong Xue	Back Pain Point	On the dorsum of the hand, midway between the transverse wrist crease and metacarpophalangeal joint, between the second and third metacarpal bones, and between the fourth and fifth metacarpal bones, 4 points in all on both hands.	Acute Back Pain/ Sprain
Yin Tang		Midway between the medial ends of the two eyebrows.	Clams the Mind/ Shen
An Mian	Peaceful Sleep	Midpoint between Yifeng (SJ 17) and Fengchi (GB 20)	Calms the Shen for Insomnia

Point	Helpful Translations	Location	Main Uses
Hua Tou Jia Ji		A group of 34 points along both sides of the spinal column, 0.5 Cun lateral to the lower border of each spinous process from the first thoracic vertebra to the fifth lumbar vertebra.	Any disorders associated with the level of the spine
Luo Zhen	Falling From Pillow	On dorsum of hand, between 2nd & 3rd metacarpal bones, 0.5 Cun posterior to metacarpophalangeal joint	Neck Pain
Shi Qi Zhuxia (Josen)	17th Vertebrae	Below the spinous process of the 5th lumbar vertebrae	Low Back Pain
Ding Chuan	Stop Asthma	0.5 Cun lateral to Dazhui (Du 14).	Stop / Reduce Asthma Attack
Ba Feng	8 Winds	On the dorsum of the foot, in the depressions on the webs between toes, proximal to the margins of the webs, eight points in all.	Foot/ Toe Pain
Ba Xie	8 Ghosts	On the dorsum of the hand, at the junction of the white and red skin of the hand webs, eight in all, making a loose fist to locate the points.	Hand / Finger Pain
Dan Nang Xue	GB Point	The tender spot 1-2 Cun below G 34.	Gall stones
He Ding	Crane Top	In the depression of the midpoint of the superior patellar border.	Knee pain/ dysfunction
Shi Mian	Lost Sleep	In the center of the heel on the bottom of the foot	Heel & Knee Pain

Point	Helpful Translations	Location	Main Uses
Si Shen Cong	Four Spirits Cleverness	A group of 4 points, at the vertex, 1 Cun respectively posterior, anterior and lateral to Du 20	Clears the Mind
Yu Yao		At the midpoint of the eyebrow.	Headache, Eye Pain
Zi Gong Xue	Uterus Point	3 Cun lateral to Ren 3	All GYN

Handling the Tools: Acupuncture Techniques

Effective acupuncture treatment consists of point selection, location and needle technique. This chapter offers an explanation of how to apply some of the different acupuncture techniques to effectively treat many common disorders.

Distal Techniques

You can often affect local pain with just distal techniques, but local treatment is usually necessary as well. In the section on treating specific conditions I will cover many useful distal treatment points and techniques.

Getting The Qi

When you're is performing acupuncture it is important to "Get the Qi." But what does that mean? According to the prevalent teachings in China today, it means that the patient must feel a deep, aching, distending, electrical or traveling sensation. This doesn't mean pain. The Qi sensation should be clear and strong to the patient, but not painful. When the patient reports the feelings associated with "getting the qi," the practitioner can stop stimulation of the point and rest assured that she has contacted the Qi. A painful acupuncture treatment is seldom necessary.

There are, however plenty of other traditions of acupuncture where the practitioner does not depend on the patient's report of feeling the Qi to ascertain that the Qi has been contacted. In fact, it could be easily argued that it is more important that the practitioner feel the Qi than the patient, since the practitioner is presumably more familiar with the feeling of Qi than the patient, and the practitioner knows better what to look for. But, how does the practitioner know what to feel, to be sure that she has contacted the Qi?

After 30 year of practice, I am pretty sure that I can feel the contact with the Qi, most of the time. Unfortunately, I can't describe it in sufficient detail, in this book, to make it clear. I can however, sometimes demonstrate it to students, when we can be together. It is a skill that can be learned but cannot be easily taught. I'm sure you will find that this skill develops, as you gain more experience. Therefore, I suggest that, until and unless you know for sure, that you have developed this skill, you should ask the patient for feedback as to when you've contacted the Qi.

The Treatment of Pain

As mentioned above, there are three important facets to performing effective acupuncture treatment of pain: choosing the correct points, precise point location, the technique applied to the points.

Understanding The Problem

First, don't get too complicated. A complicated diagnosis can often distract from a successful treatment. Most pain for example, whether it is back pain, neck pain or leg pain; whether it is from a musculoskeletal injury, arthritis, bursitis, an inflammatory reaction, or as part of a sequela of stroke; will manifest as obstruction of the meridians due to either Qi (energy) Stagnation or Xue (blood) Stagnation. When treating the pain with acupuncture and related techniques, it is often of little consequence to the treatment, whether it is from Qi or Xue Stagnation. The main difference between these two syndromes lies more in the prognosis rather than treatment approach. The treatment, with acupuncture and associated techniques, is usually the same. However, in general, a more aggressive approach is used for Xue stagnation than is used for Qi stagnation. This differentiation also carries more significance when choosing topical or internal application of herbs,. The complications of Cold, Heat, Dampness, and Wind can also often influence the treatment, but this is secondary to moving the stagnation.

Immobilization

A commonly overlooked point in the treatment of injuries is immobilization. When there is a soft tissue injury, such as a strain or sprain, it is important to immobilize the area as much as possible until the discomfort has been relieved. When injured, the body's response is to produce pain and swelling. This mechanism serves to protect the area and allows for natural healing to occur. In our modern society, we want to minimize down time and get back to work (or play) as soon as possible. So we reduce swelling, take pain relievers, or acupuncture and herbal wraps, and get right "back on our feet." This approach may fit some societal needs, but it may not the best approach for one's physical needs. Injuries take time to heal, and they require rest too. Many cases of chronic pain syndromes originally began with Qi and Xue stagnation, which stem from improper treatment of soft tissue injuries *i.e.* using the injured part too soon. I always recommend that soft tissue injuries are well splinted, or wrapped, and that patients avoid using the affected part as much as possible, until most all the pain is gone. This leads to less trouble with Cold, Dampness, Stagnation, and the like, getting trapped in the meridians and causing future problems.

Ice & Heat

The problem of cold getting trapped brings up the question of the application of heat or cold to an injury. In acupuncture theory it is well established that cold can penetrate an injured joint and stagnate, causing additional pain and possibly long term problems. I usually recommend using heat (moxa or TDP lamp) on injuries as soon as it is feasible, to move stagnation.

Use The Appropriate Technique

The question that may surface when considering the treatment might be, "Do I need to see immediate change from the treatment?" Yes. There should be some change in the condition at the time of the treatment. This is not to say that the treatment is ineffective if you do not see immediate change, but immediate results are a very good indication that you have found an appropriate treatment. I suggest that you continue to try different approaches to the problem until you get immediate results. This is not always possible, and the patient will often find improvement after some time, but it is preferable to get those "right-before-your-eyes" changes. So keep trying different techniques until you hit the one that clicks for that particular patient, at that particular time. While we are on the subject, consider that a strong acupuncture treatment for painful condition, will often leave the treated area sore, from the stimulation. This soreness should resolve in 12-24 hours. My brother tells all of his patients to wait 24 hours before judging the effect of a treatment. I think this is wise advice.

Palpation

It would be difficult to overstate the usefulness of palpation. I have seen many practitioners, some of whom I admire and respect, practicing with little or no palpation. While they may get good results, I don't understand how they know what to treat. To me it is vitally important to know exactly where the Qi is, where it is not, where it flows smoothly, and where it is stuck. Simple palpation of the meridians, acupoints, and trigger points, can easily reveal much of this information.

In order to locate a trigger point, it is usually best to palpate muscles with deep cross-fiber palpation. Once a trigger point is located, be sure you keep track of it with your fingers, so you can be sure to accurately treat it with the needle.

This is what I mean by precise point location. Be sure you've located the point of disharmony before you place the needle.

Local Treatment

Should one treat the local area or avoid it? This is often debated amongst practitioners. I say treat it, in most cases. I hedge my bets here, because the practitioner must determine what will make the problem worse and what will make it better. If, for example, the patient reports that massage makes the problem worse, or if previous treatment to the area has resulted in an exacerbation of the problem these could be indications that direct treatment may be cautioned, or contraindicated. However, I seldom see direct treatment exacerbate a painful condition, and even in those occasional situations, a few days of respite from treatment has allowed the condition to return to its pre-treatment level of discomfort, or better. The caveat here is that some problems benefit more from local, adjacent, and distal treatment, while others respond better to distal treatment only. How to asses the difference prior to treatment is not always clear. You'll have to trust your experience and intuition.

Moving Stagnation_____
In order to alleviate pain, the stagnation needs to be moved. While this is obvious, it must be consistently focused upon. Whether it is Qi or Xue stagnation, often the best course of action is simply to move it.

I believe that moving stagnation takes precedence over building deficiencies, when treating pain, in almost every case.

This is an important treatment strategy. If one attempts to build deficiencies in the presence of stagnation, it can lead to increased stagnation, hence more pain, or pain that is more difficult to treat. I have often found that. once the stagnation is successfully resolved, then deficiencies can be tonified. If the patient is extremely deficient, there may be cause for concern, but most of the time the deficiency can be better addressed after the pain has been treated. Most individuals will easily tolerate a bit of a decrease in Qi if it results in significant pain relief. Remember that pain itself affects an individual's Qi. Relieving the pain with a dispersing treatment can make the patient feel better, and often stronger as well. My wife asserts that she regularly feels a deficient patient's weak pulses become stronger after strong Qi-moving therapies, like cupping. She believes this may be because Qi that was previously stagnant becomes available to the body, increasing the overall Qi of the body.

Treating Pain with Local, Adjacent, & Distal Points_____

This is the standard approach for Treating Pain. The first question I look to answer is: "Where is the primary disharmony?" This is may seem obvious and, keeping in mind what I said earlier about not getting too complex, there is usually some looking to do. Diagnosis is like being a detective of sorts. One must find a number of clues before coming to a conclusion. If the problem is orthopedic in nature, the primary disharmony is usually at or near the site of the pain. When the primary disharmony exists someplace other than at the site of the pain, it is often found proximally. How does one assess if the primary disharmony is at the pain site or proximally, or distally for that matter? Palpation. I recommend that the practitioner look at the usual trigger points that are associated with the area of pain. (Get yourself a trigger point chart, if you don't know them.) Then explore the local and adjacent musculature. If that isn't sufficient, follow the dermatomes to the area on the back that corresponds to the more distal or more anterior pain, and then palpate. These dermatomes do not need to adhered to 100%. Rather, one should have an understanding of the pathways of the dermatomes and how they relate to the 12 main meridians and their corresponding Sinew (tendino-muscular) meridians and integrate the information gained from palpation to establish the most likely candidate areas for treatment.

All three of the following point types should be explored when treating pain.

Distal means you should use points that are distal to the pain and which affect the area of the pain. For instance:

> *Bleeding a Jing point on the effected meridian (**and/or**)*
> *Using LI 4 and Lu 7 for Pain in the head and/or neck (**and/or**)*
> *Opening the Posterior Zone with SI 3 & UB 62 for back pain*

Local means using points at the pain site: Trigger Points. It's sometimes half-jokingly put this way: "If it hurts, put a needle in it." I don't mean to say that trigger points are the most important. The practitioner is well served by being conscious of the acupuncture points, and established trigger points and their proper location. The proper location is like using a map. A map shows you where the place is, but it is only approximate. The location indicated is only as good as the map. And the map is only a symbol. So the best practice is to know the location of the local acupuncture points and established trigger points to get you to the area where the Qi can be most easily accessed, and then palpate for the exact location of the point to be needled.

Adjacent means to palpate - radiating out from the painful site -for adjacent points that are reactive. In short, finding the associated trigger points. Some sources consider related points at adjacent joints to be adjacent points. For example, using LI 10 for shoulder pain. I have no problem with this, but it's not a substitute for trigger point palpation.

It's not a bad idea to choose local, adjacent, and distal points from standard points on the Yang meridians, whenever possible. The Yang meridians are best for treating Yang disorders, and most musculoskeletal and neurological pain syndromes are generally considered Yang disorders.

Major Yang Meridian Points to keep in the forefront of your mind, for easy and regular application as local, distal and adjacent points:

> LI 1, 4, 10, 14, 15, 20
> ST 6, 7, 30, 36, 44, 45
> SJ 1, 3, 5
> GB 2, 8, 14, 20, 21, 26, 34, 41, 44
> SI 1, 3, 11, 18
> UB 1, 2, 10, 13, 18, 23, 40, 60, 62, 67
> DU 4, 14, 20
> LR 1, 3, 13 (The Liver is the Exception the the Rule. Liver

points work as well as Yang Meridian points, for treating pain.)

Sinew (Tendino-Muscular) Meridians_____

Keep the Pathways of the Yang Sinew Meridians in mind, and Treat Jing Points to clear pain from them.

Zonal Treatment for Treating Pain by Area_____

The following zones can be opened or activated by treating the distal points associated with them. **Opening these zones is one of the best ways to begin a treatment for pain**.

> **Posterior Zone: SI 3, UB 62**
> **Lateral Zone / Low Back/Hip/Sciatica: GB 41, SJ 5**
> **Anterior (Internal) Zone: Lu7, KD 6**
> **Anterior (External) Zone: LI4 - ST 36**
> **Medial (Internal) Zone: Sp 4, PC 6**
> **Medial (External) Zone: PC 6 – LR 3**

Zonal Treatment Procedure:
   1. Determine in which Zone the disorder is manifesting
   2. Treat appropriate zonal point that lies closest to the pain
   3. Treat the second associated zonal point on the opposite side
   4. Proceed with the rest of the treatment (Local, Distal, Adjacent, Microsystems, etc.)

Example: <u>Pain in Right Buttock</u>
   1. Posterior Zone, (Lower Right Quadrant)
   2. Needle: <u>UB 62 on the Right</u> (Posterior Zone, Right Side, Lower Body)
   3. Needle: <u>SI 3 on the Left</u>
   4. Palpate, and treat Local and Trigger Points as necessary, etc.

This is a simple procedure that has a big pay off.

Microsystems

I use points from the Ear, and Scalp in most all treatments for pain. While I have not covered them in this book, microsystems are very useful, and one should consider using them for all painful conditions. Microsystem points are particularly helpful in that, when treated correctly, they often have instant effects. Hence, you get clear and direct feedback on the effectiveness of your technique and diagnosis. You should know a good selection of points and lines from the following microsystems of acupuncture: **Auricular Acupuncture, Scalp Acupuncture Points & Lines, and Chinese Hand Acupuncture**. There are many good charts for finding these points.

The indications for most all Microsystem points are apparent from their names. For example, the foot point on the ear is good for treating any problems associated with the foot. Perhaps in the next edition I will cover them in some detail.

Scalp Acupuncture can be especially effective, and when done correctly, it it's no less comfortable than other forms of acupuncture. If you haven't learned APS-USA (AcuPractice Seminars Unified Scalp Acupuncture), or YNSA (Yamamoto New Scalp Acupuncture), then I recommend that you learn one or the other. You'll be glad you did.

The next page lists a group of points that are the main local points on various areas of the body, that are particularly useful for treating pain. All practitioners should be thoroughly familiar with them. They are points that you will likely be using over and over again in your practice.

The Main Local Points for Treating Pain

Head

St 3, 5, 6, 7, 8	LI 20
GB 8, 14, 20	DU (GV) 20
SI 18	

Neck

GB 20, 21	UB 10
DU 14	LI 17 area

Shoulder

GB 21	LI 14, 15, 16
SJ 14, 15	SI 9,10,11,12

Jian Nei Ling (½ way between LI 15 and the superior anterior end of the axial crease)

Elbow

LI 11,12	SI 8
SJ 10	

Wrist

LI4, 5	SJ 4
SI 3, 4, 5, 6	

Hip

GB 26,27,29,30

Knee

St 35, 36	GB 34
SP 9 (Yin)	

Ankle

St 41	GB 40
UB 60, 61, 62	

The Recipes:
Treating Specific Painful Conditions

How To Use These Recipes

First, I will often begin a course of treatments by bleeding 8-10 drops from the Jing point(s) on the affected meridian(s). I usually find that this works well for the first treatment or two, and can substantially reduce pain and discomfort. Subsequent treatment of Jing points generally offers only marginal results.

Secondly, I usually open the appropriate Zone.

The next step I'm likely to take is to add in other distal points, chosen from the recipes below.

Then I'll likely proceed with local and adjacent treatment as determined by palpation, informed by my knowledge of the appropriate acupuncture points and trigger points and their referral patterns.

While this is a good approach in many, if not most cases, it won't work every time. If it doesn't, it's time to look more carefully at the particular patient and pattern, to determine the best course of treatment.

Headaches
Note: During a headache, use only light/gentle stimulus on local points

General Headache Points: Use these points for all headaches, add specific points (below) for specific headaches

LI 4, LU 7, SJ 5, GB 41, LR 3, Ear Shen Men

Frontal Headache: St 36, St 8, GB 14, UB 2, Yu Yao

One-Sided Headache: Tai Yang, GB 8 on the affected side

Eye Headache / Pain: UB 2, GB 2 on affected side

Occipital Headache: GB 20

Vertex Headache: Du 20, Si Shen Cong

Of course many headaches are chronic, and may require a deeper look into the pattern of disharmony that is causing the headaches. Still, these treatments work well in many cases.

Facial Pain
TMJ and/or Tooth Pain: LI 4, LU 7, ST 44,

Add 1 or more of the following based on pain location:
St 3, St 4, St 5, St 6, St 7, SI 18
Scalp: **Lower 2/5th of the Sensory Area
on the opposite side**

Trigeminal Neuralgia: Same as above: Use caution on same side of pain. Often it's best to focus on distal points and use local points on the opposite side from the pain, to avoid triggering the pain.

Neck Pain & Stiffness

"Nape & Neck" and "Shoulder" Points on the Hand (located in both proximal metaphyses of the metacarpophalangeal joint of the index finger) should always be tried when treating neck pain & stiffness.

Anterior Neck Pain/ Stiffness

LI 4, LU 7, ST 36 Palpate and treat as necessary, local and adjacent trigger points, especially along the SCM and Scalene, following down to the pectoral region.

Lateral Neck Pain / Stiffness

LI 4, LU 7, SJ 5, GB 41, Palpate and treat as necessary, local and adjacent trigger points especially along the SCM and Scalene.

Posterior Neck Pain / Stiffness

SI 3, UB 62, Palpate and treat as necessary, local and adjacent trigger points, especially along the trapezius and following tight muscles down into the back. The trigger points often cross the midline in the upper, mid and lower back, so be sure to palpate both sides of the spine.

Wrist & Hand Pain

This is often associated with trigger points in the forearm. Palpate and treat as necessary, local and adjacent trigger points starting at the elbow and working your way down to the wrist. Of course you can and should treat local points, especially choosing from the effected Meridians. Perhaps choosing from the following:

LI 4, 5:	**LU** 7, 8, 9, 10	**SJ** 3, 4
SI 2, 3, 4, 5, 6	**PC** 7, 8	

Arm & Shoulder Pain

Anterior Arm/Shoulder Pain

LI 4, ST 36, LI 14, LI 15, Jian Nei Ling Palpate and treat as necessary, local and adjacent trigger points.

LU 7 and KD 6 might be substituted for LI 4 & St 36, if the pain is closer to the LU meridian, or if the first set of points are not effective.

Lateral Arm/Shoulder Pain

SJ 5, GB 41, LI 14, LI 15, GB 21, Palpate and treat as necessary, local and adjacent trigger points.

Posterior Arm/Shoulder Pain

SI 3, UB 62, SI 9 and/or SI 10, SI 11 and/or SI 12, Palpate and treat as necessary, local and adjacent trigger points.

Upper and Mid Back Pain

SI 3, UB62 to open the Posterior Zone

There are a number of good acupuncture points for treating the mid and upper back, including the traps, but the **Treatments of choice for this area are <u>Cupping and Gua Sha</u>.** Cupping and Gua Sha can often release stagnation in this area quite effectively, and should usually be part of the treatment approach.

Trigger points primarily, especially in the region of:
SI 11, and GB 21.

Also chose points from:
Du 8 -14 and the associated Hua To Jia Ji Points

Low Back & Hip Pain and Sciatica

There are many influences on the low back and hips. Consider that the following Meridians all pass through and influence the lower back:

Urinary Bladder
Du Mai (Governing Vessel)
Dai Mai (Belt Vessel)
Gall Bladder

In addition the strength of the low back is dependent on the vitality of the Kidneys.

Given all that, I have found that **Opening the Dai Mai with GB 41 & SJ 5** can be a highly effective treatment. That's how I often begin a course of treatment. If you try this approach, you will be surprised at how often it gives very good results.

Another very good approach is to begin by **Opening the Posterior Zone, with SI 3 & UB 62.**

The following points all are useful local points for treating the low back and hips, and should be considered for treatment. Along with the usual palpation and treatment of Ashi and Trigger Points.

GB 26, Du 4, UB 23, UB 52, UB 25, UB 32.

Distal Points include **UB 40, UB 60, UB 67**

Knee Pain

Knees can be a little stubborn for a number of reasons. First, they take a lot of abuse from bearing excess weight, suffering from improper foot position when walking, and the relatively Yin location of the knees. Remember Yin is substantial, and the further down in the body one goes, the more substantial the problem becomes, and subsequently the tougher to move. That said, there are some good points to treat the knees.

Distally, in addition to the usual Jing points and Zonal points, you can stimulate Shi Mian (center of the heel). I usually use moxa for this, but you can also needle it.

Adjacent treatment generally consists of palpation above and below the knee, especially focusing on the calf to find trigger points.

Local Points: ST 36, 35, 34, GB 34, Sp 9, 10, Xi Yan, UB 40

Foot & Ankle Pain

Once again, the Yin location of the feet sometimes makes them difficult to treat, so I recommend you use a lot of points on the feet. I feel patients benefit from even 10 -12 points on each foot, if needed. I mainly choose from the Yang meridians, but don't limit myself to them. The Liver meridian can be very useful here too. Local points to choose from include:

    Check the calf carefully for trigger points

    SP 6 and GB 39 together to open the meridians in the foot & Ankle

    Any of the Jing Points: UB 67, GB 44, ST 45, LR 1, Sp 1,

    ST 44, 42, 41

    GB 43, 42, 40

    UB 60, 61, 62, 63, 64, 65

    LR 2, 3, 4

SP 2, 3, 4, 5
KD 2, 3, 4, 5, 6

Abdominal Pain

Abdominal Pain is clearly an internal medical problem and there are many reasons for it. With so many diagnosis resulting in abdominal pain, it's difficult to offer comprehensive treatment advice in this work. However, there are a few treatments that can often help. As with all disorders, be sure you've diagnosed the patient carefully before beginning treatment.

Open the **Dai Mai GB 41, SJ 5**
　　　And/Or
Open the **Chong Mai SP 4, PC 6**

Choose from: **Ren 2, 4, 6, ST 25, REN 12**

Internal Disorders

When we use the term "Organ" or the names of the Organs, in TCM, we do not infer the identical meaning as we understand them in current biological medicine in the West. We use these terms to address a complex system of interrelationships that, while they include the organs as we understand them, they also include emotions, thoughts, and other physical systems in the body, or body/mind/spirit. In this way the ancients in China could understand and work with all the aspects of an individual within the context of just a handful of systems. While modern biology cannot fully explain this approach, there are countless generations of experience to attest to its efficacy.

Each Organ has a specific job to perform. I will discuss the main jobs of each Organ, and the typical symptoms associated with their primary dysfunctions. There are, of course, many other associations for each Organ, but I'll only address those associations that have clear clinical application, and are regularly encountered in daily practice. I am focusing on the functions of the Yin Organs, since they are the ones that perform most of the jobs that I address in this book. For the most part the Yang organs support the Yin Organs in their operation. The only Yang Organ I'll address here is the Stomach.

I have focused this edition of this book on the treatment of painful conditions. However, I did want to include at least a little direction on internal medicine. Perhaps in the next edition, I will give more detailed "recipes" for internal disorders. For now, I hope you and your patients find this, admittedly limited, set of directions helpful.

The Liver

The Liver is responsible for the smooth and easy flow of energy and emotions in an individual. This includes assisting all processes, physical, mental and emotional in flowing smoothly and regularly. Muscle contraction, menstrual cycles, and managing the smooth flow of the emotions--especially anger, are just a few examples of this function.

When the Liver is in disharmony, the following are some common possible symptoms:

Erratic mood swings / Moodiness / Irritability
Quick to Anger /Volatile-violent outbursts
Frustration
Compulsive energy
Excessive muscular tension
Throat clearing/ feeling of plum pit in the throat
Neck and Shoulder Tension
Pre Menstrual Syndrome with breast pain and distention
Painful Menstrual Cycle with clots and sharp stabbing pain
Inguinal pain and Hernias
Contractures, spasms / uncoordinated movements
Acute inflammatory problems: i.e. Herpes, Conjunctivitis
Eye problems: Tearing, blurry vision, Night Blindness,
Floaters, photophobia & light sensitivity

Treating The Liver

Acupuncture is quite good for soothing the Liver, but the patient will have to make some lifestyle changes to keep the condition from returning. Some type of stress reduction technique needs to be employed by the patient. Usually people need to understand that rest, relaxation, and repose are as important as, or more important than, achieving their external goals. I know it's a tall order, but how else can you get folks to let go of vain strivings that are affecting their health?

Patients should consider doing some gentle form of yoga, or tai chi or Qi gong. The gentle stretching of the tendons is a wonderful way to relax the Liver, and it usually helps with the stress too. (GENTLE stretching, not strenuous, hot, & sweaty types of yoga).

One other thing is removing coffee from the diet. Coffee (not caffeine) has a particular affinity for the Liver, and I believe it contributes greatly to Liver Qi stagnation.

To sum up:
Acupuncture once a week to move LR Qi.
LR 3, GB 41, PC 6, SJ 5
Gentle Yoga
Quit Coffee
1-3 months produces excellent results in most cases

Before we move on to the Spleen let me address IBS. I've had mostly good results with IBS, as it is often a result of Liver overacting on Spleen. If the patient fits this dx, harmonizing the Liver, will allow the Spleen to come into balance. This assumes the diet is reasonable.

Treat the Liver (Above) and Tonify the Spleen (Below).

The Spleen

The Spleen is responsible for managing all aspects of Digestion and Assimilation, Transformation and Transportation. This is mainly seen as managing the transformation of food and fluids into energy, blood, body fluids and tissues. When the Spleen is not functioning well, food and fluids are not fully processed, one's energy level drops, and digestive problems become apparent. These unprocessed products then accumulate in the individual as excess weight and phlegm and mucous.

When the Spleen is in disharmony, the following are some common possible symptoms:

> Fatigue
> All problems with the Lower Digestive Processes
> Overweight/underweight
> Bloating, Gas
> Loose Stools
> Prolapses
> Mucous in Stools
> Blood in Stools
> Excessive or insufficient Menstrual bleeding
> Excessive Bruising
> Vomit
> Hemorrhoids
> Muscle weakness/atrophy

Treating The Spleen

Since the Spleen mainly suffers from deficiencies of Qi, Yang, and Blood, it tends to respond best to Diet and Herbal Treatment. However, acupuncture can often be very helpful and is certainly worth a good try, especially if you can add some moxa for tonification. Also Tonifying the Kidneys is often helpful when treating the Spleen. Keep in mind that tonifying the Spleen can take some months.

When treating the Spleen, you will generally want to tonify all the points:

ST 36, SP 6, Ren 6, Ren 8 (moxa only), UB 20

Adjust Diet: Cut out or reduce highly processed, cold and damp foods

The Stomach

The Stomach is the one Yang Organ we will consider in this chapter because it is so important in the early digestive process. While the Spleen is responsible for all digestive processes, once the food has left the Stomach, problems with the Stomach itself are usually addressed directly. Heartburn, hiatial hernias, and stomach ulcers are all symptoms of Stomach disharmonies.

When the Stomach is in disharmony, the following are some common possible symptoms:

      Stomach Pain / Distention

      Nausea

      Vomiting

      Belching

      Acid Regurgitation , Heartburn, Ulcers

      Bad Breath

      Increased/decreased appetite

Treating The Stomach

The Main points for treating the Stomach are the same for most conditions:

      ST 36, SP 6, Ren 12, UB 21

      If there's Heat: add ST 44 and perhaps ST 45

The Kidneys

The Kidneys are responsible for the balance of Yin and Yang in the body. They regulate our constitutional energies, and manage reproductive, urogenital, and sexual functions. Urinary, prostate, premature aging, and lowered libido are a few of the disharmonies we associate with the Kidneys. **The Kidneys are "housed" in the lower back, and so are responsible for the strength of our low back.** This is useful in treating chronic low back and sciatic disorders.

When the Kidneys are in disharmony, the following are some common possible symptoms:

Low back pain & weakness
Pain & weakness of the Legs, Knees, & Ankles
Impotence and Infertility
Incontinence
Polyuria
Loss of hearing / Tinnitus
Lower body Edema
Problems of growth and development
Signs of premature aging
Sciatica
Low sex drive
Fearfulness & Lack of Will
Fatigue

Treating The Kidneys
Like the Spleen, Treating the Kidneys most always involves tonification. Like the Stomach most Kidney treatments involve the same points:

> **KD 3, KD 16, UB 23, DU 4**
> **If there's Heat (from Yin Deficiency) add KD 2 and KD 7**

The Heart

The Heart is the Organ that stores our "Shen." The Shen is our spirit, in the sense of that aspect of the awareness which is reflected in our eyes; our sense of our "self" our perception of the world around us, and how we fit in. In a word, consciousness. Disharmonies of the Heart can manifest as any disturbance of the consciousness, from mild anxiety or depression to severe psychological disorders. All these disorders fall under the term of "Shen Disturbance."

When the Heart is in disharmony, the following are some common possible symptoms:

All disorders effecting the Mind: Spirit, Consciousness, Memory, & Thinking

Most Sleep disorders

Most Heart disorders can be effectively treated by treating the Pericardium.

The Pericardium

The Pericardium is responsible for Setting the Order of the Heart. In other words, we treat the pericardium for all the organic functioning of the heart:

Treat the pericardium for tachycardia, bradycardia, irregular heartbeat, chest pain, palpitations, etc.

The Pericardium also treats the psychological aspects (Shen) of the Heart.

Treating The Heart and Pericardium

Most all aspects of the Spirit can be treated with the **Tai Ji Treatment** discussed earlier in the book **(PC6, SJ 5, LR 3, GB 41)**

Fast, Slow or Irregular Heartbeat **PC 5, Ht 7**

HT 8 will reset the order of the heart

PC 8 for dream disturbed sleep

The Lungs

The Lungs are responsible for respiration, energy (QI) production, and protection from outside influences, similar to our concept of the immune system. When the Lungs are functioning well, the individual is vital and strong. When they are in disharmony, we see respiratory problems like COPD, shortness of breath, and a tendency towards upper respiratory infections.

When the Lungs are in disharmony, the following are some common possible symptoms:

All types of respiratory disorders
All sinus disorders
Many Skin Disorders
Common Colds and Flus
Fatigue

Treating The Lungs

Tonify the Lungs: **Lu 9, Lu 1, UB 13**
Colds and Flus: **LI 4, LU 7, LI 20,**
Cough **LU 5**
Sore Throat or other Heat in the Lungs: **LU 10**